1 MONTH OF
FREE
READING

at

www.ForgottenBooks.com

By purchasing this book you are eligible for one month membership to ForgottenBooks.com, giving you unlimited access to our entire collection of over 1,000,000 titles via our web site and mobile apps.

To claim your free month visit:
www.forgottenbooks.com/free1297127

ISBN 978-0-428-98072-6
PIBN 11297127

This book is a reproduction of an important historical work. Forgotten Books uses
state-of-the-art technology to digitally reconstruct the work, preserving the original format
whilst repairing imperfections present in the aged copy. In rare cases, an imperfection in
the original, such as a blemish or missing page, may be replicated in our edition. We do,
however, repair the vast majority of imperfections successfully; any imperfections that
remain are intentionally left to preserve the state of such historical works.

VOL.5 No. 23   JUNE 10, 1955

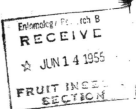

Cooperative
ECONOMIC INSECT
REPORT

Issued by

PLANT PEST CONTROL BRANCH

AGRICULTURAL RESEARCH SERVICE

UNITED STATES DEPARTMENT OF AGRICULTURE

AGRICULTURAL RESEARCH SERVICE

PLANT PEST CONTROL BRANCH

ECONOMIC INSECT SURVEY SECTION

The Cooperative Economic Insect Report is issued weekly as a service to American Agriculture. Its contents are compiled from information supplied by cooperating State, Federal, and industrial entomologists and other agricultural workers. In releasing this material the Branch serves as a clearing house and does not assume responsibility for accuracy of the material.

Reports and inquiries pertaining to this release should be mailed to.

Economic Insect Survey Section
Plant Pest Control Branch
Agricultural Research Service
United States Department of Agriculture
Washington 25, D. C.

Volume 5   June 10, 1955   No. 23

COOPERATIVE ECONOMIC INSECT REPORT

Highlights of Insect Conditions

GRASSHOPPERS remain primary insect problem in New Mexico, continue to build up in Nebraska, and threatening populations remain in Kansas. Situation severe in Missouri. Other States report infestations. (pages 505, 529, 530).

ARMYWORM damage to small grains continues in Kansas and Missouri but beginning to decrease. Infestations remain spotted in Illinois. Damage appearing in Delaware and Maryland. Outbreaks subsiding in North Carolina, over in Tennessee. (pages 506, 529, 530).

ALFALFA WEEVIL reported from North Carolina for first time. Damage continues in Virginia, Maryland, and Delaware. (page 510).

CITRUS BLACKFLY found at Brownsville, Texas. (page 514).

MEDITERRANEAN FRUIT FLY reported from Costa Rica. (page 514).

PEA APHID active on peas in several States. (page 517).

Light infestation of GREEN PEACH APHID on tobacco in some North Carolina counties. (page 519).

THRIPS continue injurious to cotton in several States. (page 524).

A SCALE (Acnidiella taxus) collected in Florida for first time. (page 522).

SOUTHERN PINE BEETLE outbreak continuing in northern Alabama. (page 520).

NOTES received too late for inclusion in the body of this issue. (pages 529, 530).

States reporting this week - 42.

****

Reports in this issue are for the week ending June 3, 1955 unless otherwise designated.

WEATHER FOR THE WEEK ENDING JUNE 6, 1955

A low pressure trough moving slowly eastward from the Rocky
Mountains to the Mississippi Valley during the week was responsible for
a northward flow of warm, moist air and frequent showers in that area.
Temperatures for the week averaged as much as 6° above normal in the
upper Great Lakes and upper Mississippi Valley and 1° to 3° above in
the Great Plains. Daily showers in north-central areas produced weekly
totals ranging from one-half to over two inches. Over the weekend
showers extended southward over the eastern portions of the Great Plains
and Mississippi Valley almost to the Gulf. Numerous severe thunderstorms
with damaging hail and wind and a few tornadoes were reported. Dry, cool
Polar air covered most of the area east of the Mississippi until the last
day of the period when the low pressure trough over the Mississippi Valley
brought showers to the area between the Mississippi and the Appalachians.
An area of low barometer off the northeastern coast was responsible for
moderate to heavy showers in New England and light showers in the
middle Atlantic States at the beginning of the period, and moderate to
heavy frontal rains fell in coastal areas of the middle Atlantic States over
the weekend. The remainder of the area east of the Appalachians received
little or no rain. Temperatures were unusually low for the season in the
Atlantic Coastal States at the beginning of the period, but had risen to
about normal levels at the end.

Unseasonably cool, dry weather continued west of the Continental Divide.
Freezing weather was rather general in central and northern interior
areas early in the week when light to moderate rain and some snow fell in
the northern half of the Rocky Mountain State area. Drought-breaking
rains in the Great Plains were the main weather feature of May. Subnormal
precipitation and drying winds during March and April in the Great Plains
had depleted soil moisture to the extent that the crop outlook was dismal
indeed. But heavy rains during the last two decades of May reversed this
outlook completely. Soil moisture is now adequate and the crop outlook
good to excellent, although the small grain crop was damaged beyond
recovery in some areas. The month of May, like April, was warmer than
normal east of the Continental Divide and unseasonably cool in the far West.
(Summary Supplied by U.S. Weather Bureau).

CEREAL AND FORAGE INsECTS

GRASSHOPPERS - PENNSYLVANIA - Nymphs abundant field margin
of alfalfa in Perry County. (Pepper). NORTH CAROLINA - Up to 25
per square yard, mainly Melanoplus mexicanus, in corn, cotton, and
other crops in Cleveland County. (Clapp, Jones). FLORIDA - First
first-instar nymphs of Schistocerca americana at Gainesville. (Kuitert).
LOUISIANA - Increased population of Melanoplus femur-rubrum and
Melanoplus sp. in and near rice fields with 4 per square foot in Acadia
Parish; 75 per 100 sweeps of alfalfa Natchitoches Parish. (Oliver).
TEXAS - Medium to heavy widespread infestations, mainly M. differen-
tialis, in pastures, field crops, legumes and fence rows in Lee, Collin,
Tarrant, Haskell, Hunt, Dallas, Ellis, Navarro, Limestone Counties
and to Oklahoma line. (Jones, Steinbach, Dahlberg, Martin, Randolph).
NEW MEXICO - Grasshoppers continue as primary insect in State.
Controls will be necessary on large areas in eastern part of State. (Ins.
Lett., May 28). CALIFORNIA - Hatching reported April 18 in Merced
County rangeland. (Cal. Coop. Ins. Rept.). MISSOURI - Population
very high, from 20 to over 200 per square yard of hatching beds. Damage
to alfalfa, soybeans and garden crops has started. Second-cutting
alfalfa heavily damaged in several areas. Grasshoppers are beginning
to spread out from hatching beds in southern half of State. (Kyd, Thomas).
ILLINOIS - Hatching of Melanoplus spp. almost complete over most of
State. In some cases grasshoppers have moved into new legume seed-
ings and are causing damage. (Moore et al). WISCONSIN - Continuing
to hatch, but hatch checked by cold wet weather. (Chambers). KANSAS -
No appreciable change in population. Threatening populations still
present and may increase. M. differentialis hatching. Counts of 30-40
Melanoplus nymphs, third and fourth instars, at survey stops in Riley
and Wabaunsee Counties. (Matthew). NEBRASKA - Populations continue to build up in many areas. From 20-25 nymphs per square yard in
alfalfa and oats in central area, 15-40 nymphs in east central, and 20-
100 in southeastern. Melanoplus mexicanus adults along roadsides in
some areas of eastern Nebraska. (Roselle, Andersen). SOUTH
DAKOTA - Up to 75-100 per square yard in Lyman County. M. mexicanus
and M. bivittatus 10-15 percent hatched, M. differentialis about 5 percent and M. femur-rubrum not yet started. (Burge, King). MONTANA -
In Golden Valley County: M. confusus third instar, Aulocara elliotti,
Ageneotettix deorum and Amphitornus coloradus first instars. M. confusus and Aeropedellus clavatus in southern area. (Twilde, Wolff).

EUROPEAN CORN BORER (Pyrausta nubilalis) - MASSACHUSETTS -
Moths active. (Crop Pest Cont. Mess.). MARYLAND - Ten percent of
wheat stems in 5-acre field broken off by small larvae. (U. Md., Ent.
Dept.). DELAWARE - Egg masses common on taller corn in southern
Kent and Sussex Counties, and on potatoes generally. Injury on corn
at Canterbury and southward. (Milliron). OHIO - Moths 50 percent

emerged at Columbus June 2. Fifteen to 20 egg masses per 100 plants on early market sweet corn. At Wooster, emergence is just starting but pupation practically 100 percent. (C. R. Neiswander). ILLINOIS - Still two weeks earlier than normal. Moth emergence in southern third of State almost complete. Pupation complete in central third, 25-50 percent of moths emerged and eggs being deposited. In northern third pupation almost completed and emergence ranges from 10-25 percent. Heavy egg deposition, 60 masses per 100 plants, in East St. Louis area. Very few fields in this area will warrant treatment. (Moore et al). NEBRASKA - Pupation 100 percent throughout corn-growing areas of State. First eggs in Pawnee County, southeastern area, June 3; 12 egg masses per 100 plants in one field of 24-inch corn (extended). Emergence in northeast district about 23 percent, central about 29 percent, and southeast 32 percent. Average corn height for State 4-6 inches. (Roselle, Andersen). NORTH DAKOTA - Pupation 50 percent June 2 in southern Cass and northern Richland Counties; some emergence. Corn height in area 3 inches. Development of borer earlier than in previous years. (Goodfellow). SOUTH DAKOTA - Some emergence in Union County and practically all borers pupated. (Lofgren). MISSOURI - Pupation and emergence about complete in all areas. Very little corn in northern half of State tall enough to be attractive for egg deposition. (Kyd, Thomas).

ARMYWORM (Pseudaletia unipuncta) - DELAWARE - Larvae, all sizes, in grass and legumes. Approximately 20 acres wheat at Harrington showing serious damage, but only light feeding on small grains in several other areas. Damage to corn also noted. (Milliron). MARYLAND - Heavy damage to barley and oats in Talbot and Dorchester Counties, and corn in St. Marys County. First outbreaks of season. (U. Md., Ent. Dept.). NORTH CAROLINA - Early instars in Buncombe County. Outbreaks in Piedmont apparently subsiding. (Jones). ILLINOIS - Infestations rather spotted in triangular area from St. Louis to Charleston, north to Champaign and westward to Peoria. As high as 32 per linear foot in exceptionally rank grains and grasses. From second to fifth instars in many fields. Parasitism still very low. (Moore et al). MISSOURI - Some damage continues in central third of State although activity slowed some. Heads of barley still being cut and increasing injury to wheat may be expected. Undercover crops of legumes and grasses in small grains heavily damaged or destroyed in many fields. Parasites destroying considerable number of larger larvae and with warmer weather expected to largely eliminate problem. (Kyd, Thomas). KANSAS - Destructive populations continue in many barley fields of eastern area. Larvae also in wheat and brome fields and some pastures. Populations decreasing in southern area, pupation underway and parasites and predators numerous. Additional loss of barley may occur in east central counties. (Matthew).

MORMON CRICKET (Anabrus simplex) - MONTANA - Third and fourth instars at lower elevations in Big Horn and adjacent counties; second and third at higher elevations. (Roemhild). WYOMING - Five thousand acres of rangeland baited southeast of Sundance. (Spackman).

CHINCH BUG (Blissus leucopterus) - ILLINOIS - Still present in many thin stands of oats. Egg-laying continues even though many eggs have hatched. (Moore et al). SOUTH CAROLINA - Reported damaging corn in many Piedmont and Coastal Plain areas. (Nettles). NEBRASKA - Attacking volunteer corn in large numbers in Pawnee County. (Andersen).

CORN EARWORM (Heliothis armigera) - NORTH CAROLINA - Two to 4 early instars per plant on corn in Harnett County, 50 percent of plants infested. (Ammons, Jones). Late instars damaging corn in Duplin County. (Brett). Countywide in Sampson County. (Morgan). MISSOURI - Whorl injury to corn by half-grown larvae in Vernon County and a few other areas of southwest corner of State. (Kyd, Thomas). ARIZONA - Unusually abundant on alfalfa for time of year at Yuma; 1 per 10 sweeps. (Ariz. Coop. Rept.). WASHINGTON - First adult taken in light trap May 31, Yakima Valley. (Klostermeyer).

CUTWORMS - PENNSYLVANIA - Moderate on corn in Centre County. A. venerabilis involved. (Adams). DELAWARE - Numerous in alfalfa and clovers in some areas. (Milliron). TENNESSEE - Light widespread on corn and gardens in San Augustine and Sabine Counties. (Markwardt). OHIO - Agrotis ypsilon and Crymodes devastator causing damage to corn following mixed meadows in northeastern area. (C. R. Neiswander). ILLINOIS - Damage by A. ypsilon beginning in corn. Peridroma margaritosa population fairly low. (Moore et al). WISCONSIN - Serious damage in Washburn and Barron Counties. (Chambers). SOUTH DAKOTA - Corn and soybeans in southeast being damaged, especially in river bottom lands. From 10-20 percent loss in fields in Union County. (Lofgren, May 28). WASHINGTON - Severe damage to sod of bluegrass grown for seed in Yakima Valley. Strawberry root weevil (Brachyrhinus ovatus) also involved. (Klostermeyer).

ARMY CUTWORM (Chorizagrotis auxiliaris) - WASHINGTON - About 1500 acres barley and wheat sprayed in Asotin County,*and 200 acres of barley injured in Whitman County. (Brannon, Bond, Entenmann). IDAHO - Insecticides controlling outbreak in Idaho County. Pupation in fields where controls not applied. High incidence of disease observed. This disease has been identified by E. A. Steinhaus as a virus of the granulosis group and, according to him, is first known instance of a granulosis virus in C. auxiliaris. (Manis, May 28). MONTANA - Severe damage in wheat in Chouteau, Teton, Fergus, and Park Counties. Locally severe and moderate damage in Liberty, Blaine, Toole, Daniels, and Gallatin Counties. Invaded newly-seeded lawns in Glendive. (Roemhild). UTAH - Outbreaks subsiding as pupation occurring. Several

* About 250 acres of rape destroyed in Asotin County

thousand acres of alfalfa and small grains treated during May. (Knowlton).

FALL ARMYWORM (Laphygma frugiperda) - FLORIDA - Heavy injury on grass at Belle Glade. Mocis sp. also involved. (Genung). LOUISIANA - About one per stalk on seedling corn in East Baton Rouge Parish. Controls initiated. (Oliver).

PALE WESTERN CUTWORM (Agrotis orthogonia) - MONTANA - Major damage in widely scattered spots over State. Thousands of acres reseeded or treated. Infestations in Yellowstone, Big Horn, Broadwater, Gallatin and Liberty Counties. Additional counties also reporting.. Damage mostly in winter wheat but barley and spring wheat damaged in some areas. (Roemhild).

SOD WEBWORMS - PENNSYLVANIA - Several corn fields with heavy infestation in Perry County. (Pepper). Fifteen acres corn completely destroyed Somerset County. (Udine).

STALK BORER (Papaipema nebris) - MISSOURI - Light damage to margins of corn, wheat and gardens in scattered areas. (Kyd, Thomas).

LESSER CORNSTALK BORER (Elasmopalpus lignosellus) - FLORIDA - Several fields of young cane show 60-75 percent plants with dead hearts at Belle Glade. (Questel).

CORN FLEA BEETLE (Chaetocnema pulicaria) - DELAWARE - Bacterial wilt appearing on corn. (Milliron).

FALSE WIREWORMS (Eleodes spp.) - KANSAS - Numerous E. suturalis beetles appearing in central area. Other species in numbers in most of western Kansas. Much greater numbers than since 1948. General buildup result of dry weather of past few years. (Matthew). MONTANA - Damage to grains very severe some areas. Abundant moisture has kept feeding at ground surface. Seed treatment has not controlled these pests. (Roemhild).

SUGARCANE BEETLE (Euetheola rugiceps) - NORTH CAROLINA - Damage to corn in Camden and Hertford Counties. (Scott). SOUTH CAROLINA - Damage to corn continues in many areas. (Nettles). KENTUCKY - Considerable numbers collected at Lexington and Louisville. (Price). MISSISSIPPI - Total of 500 specimens collected in one light trap in one night. Damaging corn in Lauderdale County; 75 percent of plants killed in one 7-acre field in Harrison County. (Hunsucker, Hutchins, McGehee).

MAIZE BILLBUG (Calendra maidis) - NEBRASKA - Damage beginning to show in corn; 9-10 plants per hundred show feeding holes. (Andersen).

WIREWORMS - IDAHO - Severe losses in winter wheat in Owyhee County.
Areas up to one acre 100 percent destroyed. (Walz, May 28). MONTANA-
Damage in winter wheat in Flathead, Pondera, Daniels, Carter, and
Fallon Counties. (Roemhild). WYOMING - Up to 5 larvae per kernel
of corn in Goshen County. (Robb). MISSOURI - Light damage, 1-4 percent,
in most fields of corn in northeast area. (Kyd, Thomas). WISCONSIN -
Considerable damage to corn and truck crops in southern area. (Chambers).

A MAY BEETLE (Phyllophaga cribrosa) - TEXAS - Heavy widespread
infestations in pastures and grasslands in Denton County. (Randolph).

EUROPEAN WHEAT STEM SAWFLY (Cephus pygmaeus) - DELAWARE -
Noticeable damage to barley near Kenton. (Milliron).

CORN BLOTCH LEAF MINER (Agromyza parvicornis) - DELAWARE -
Adults prevalent in Kent and Sussex Counties. Larval damage to lower
corn leaves common some areas. (Milliron).

SEED-CORN MAGGOT (Hylemya cilicrura) - WISCONSIN - Damage to
corn in low ground in muck areas. (Chambers).

HESSIAN FLY (Phytophaga destructor) - MISSOURI - Lodging of wheat
common throughout State, up to 17 percent of stand. (Kyd, Thomas).

RICE STINK BUG (Solubea pugnax) - LOUISIANA - Ten to 16 per 100
sweeps in grass and rice in Acadia Parish. (Oliver).

ENGLISH GRAIN APHID (Macrosiphum granarium) - CALIFORNIA -
Generally distributed in Santa Barbara and San Luis Obispo Counties.
Heavy in some wheat fields. Also on barley and grasses. (Cal. Coop.
Rept.). KANSAS - Moderate to heavy infestations continue to cause concern in wheat and barley in southeast and east central areas. From
1-40 aphids per head in many fields but not all heads infested. (Matthew).
NEBRASKA - From 20-25 per 25 sweeps on wheat in eastern and southeastern areas. (Andersen). MISSOURI - Populations declining in wheat
in most of central third of State. Spring-seed oats becoming heavily
infested in several areas, 6-35 per head. (Kyd, Thomas). ILLINOIS -
Infestation on wheat heads very variable in southwestern area. Very
little treatment has been needed or will be made as wheat nearing
maturity. (Moore et al).

EUROPEAN CHAFER (Amphimallon majalis) - NEW YORK - Nine adults
taken in trap in Geneva area June 2, earliest observed date of flight.
Heavy flight may be expected after June 15 in Newark, New York area.
(Tashiro).

SWEETCLOVER WEEVIL (Sitona cylindricollis) - IDAHO - Injury in
all sweetclover in southwest area. (Walz).

ALFALFA WEEVIL (Hypera postica) - DELAWARE - Heavy larval
and adult feeding retarding second-growth alfalfa throughout State.
(Milliron). MARYLAND - Cocoons numerous in alfalfa fields in Harford
County, very little damage to second cutting. Second-growth alfalfa
damage by larvae in Worcester County. (U. Md., Ent. Dept.).
VIRGINIA - Larvae and adults continue to heavily damage alfalfa in
many eastern counties, May 27. (Morris). NORTH CAROLINA -
Infestation found on alfalfa in Granville and Vance Counties. Slightly
over 6 larvae per 100 sweeps. (Dogger). According to ARS files, this
is the first record of this insect in North Carolina. PA. - Found in
alfalfa in Carbon, Monroe, Juniata, Snyder, Pike, Northampton, Lehigh,
Luzerne and Northumberland Counties. (App, Negley, Menusan).

LESSER CLOVER LEAF WEEVIL (Hypera nigrirostris) - NEBRASKA -
Heavy in east and southeast; average of 19 of 25 stems in various fields
showing injury. From 9-18 adults per 25 sweeps. (Andersen).

POTATO LEAFHOPPER (Empoasca fabae) - MARYLAND - Five adults
per 10 sweeps in alfalfa field Worcester County, first of season. (U. Md.,
Ent. Dept.).

THREE-CORNERED ALFALFA HOPPER (Spississtilus festinus) -
LOUISIANA - Increasing in alfalfa. Per 100 sweeps in following parishes:
Rapides 47, Natchitoches 55, Bossier 18. (Oliver).

SPITTLEBUGS - PENNSYLVANIA - Adults becoming abundant on legume
hay in south central area. (Pepper). DELAWARE - Continues to injure
alfalfa in Middletown area. (Milliron). MARYLAND - Meadow spittlebug
adults numerous in alfalfa and clover in most sections. (U. Md., Ent.
Dept.). MICHIGAN - Philaenus leucophthalmus numerous at Kalamazoo.
(Hutson).

YELLOW CLOVER APHID (Myzocallis trifolii) - ARIZONA - Apparently
checked in southwestern counties, also in Bowie area of Cochise County.
Population drop correlated with abundance of convergent lady beetle.
Still severe in Greenlee County. (Ariz. Coop. Rept.). NEW MEXICO -
Continues of extreme importance. Second cutting of alfalfa in southern
half being heavily attacked with every field in Mesilla Valley showing
populations. Large populations at first cutting in Fort Sumner area.
(Ins. Lett., May 28). KANSAS - Threatening infestations in some
alfalfa fields in south central area have diminished and not problem in
that area now. Parasites and predators have built up and, with weather
changes, are giving some control. Aphids still being found in many
alfalfa fields in that area, however. (Matthew). OKLAHOMA - Twenty-
five per 125 sweeps, about same as last week, in one alfalfa field in
Stillwater area. (Fenton). PENNSYLVANIA - Found on red clover in
Northumberland County. (App).

PEA APHID (Macrcsiphum pisi) - PENNSYLVANIA - Very heavy on alfalfa in Lehigh and Northampton Counties. (Menusan, Negley). MARYLAND - Damage to second-growth alfalfa in Harford and Worcester Counties. (U. Md., Ent. Dept.). SOUTH CAROLINA - Damage to alfalfa in Greenville, Laurens and Abbeville Counties. (Nettles). MISSOURI - Building on new-growth alfalfa following first cutting, from 8-25 aphids per sweep. (Kyd, Thomas). CAL. - Heavier than usual Merced and Stanislaus Counties. (Cal. Coop. Rept.)

APPLE GRAIN APHID (Rhopalosiphum fitchii) - MICHIGAN - Numerous on oats at Kalamazoo. (Hutson).

THRIPS - NEW MEXICO - Large populations continue on alfalfa, onions, and cotton. (Ins. Lett., May 28).

GARDEN WEBWORM (Loxostege similalis) - SOUTH DAKOTA - Moths very abundant in eastern area for past two weeks. (Lofgren, May 28).

BEAN LEAF BEETLE (Cerotoma trifurcata) - DELAWARE - Destructive to soybeans in some areas. (Milliron). ILLINOIS - Feeding extensively on newly-emerging soybeans. (Moore et al).

VETCH BRUCHID (Bruchus brachialis) - MICHIGAN - Numerous in Ingham County. Entering fields from hibernation. (Hutson).

CLOVER MITE (Bryobia praetiosa) - IDAHO - Abundant around homes in Moscow. Considerable damage to grass and clover in lawns, and severely damaging many flowers and shrubs. (Manis, May 28).

WHITE-LINED SPHINX (Celerio lineata) - TEXAS - Light widespread on vetch in Kaufman County. (Randolph). MISSOURI - Small areas of Vernon County and other widely scattered spots of southwest have very heavy numbers of larvae. Apparently feeding on a species of purslane. (Kyd, Thomas).

PLANT BUGS - ARIZONA - Very abundant, 700 per 100 sweeps, in one field of alfalfa at Buckeye, Maricopa County. Much lower than usual on alfalfa at Yuma. (Ariz. Coop. Rept.). MONTANA - Local severe infestations of Labops hesperis in crested wheatgrass and barley in Stillwater and Yellowstone Counties. (Roemhild). LOUISIANA - Per 100 sweeps in alfalfa by parish: 53 adults, 31 nymphs in Rapides; 38 adults, 23 nymphs in Natchitoches; and 6 adults and 5 nymphs in Bossier. (Oliver). PENNSYLVANIA - Adults becoming abundant on legume hay in south central area. (Pepper).

A LEAF MINER (Liriomyza sp.) - ARIZONA - Damage quite prevalent on older leaves of alfalfa in Buckeye area. Adults abundant at Yuma. (Ariz. Coop. Rept.).

FRUIT INSECTS

CODLING MOTH (Carpocapsa pomonella) - MASSACHUSETTS - Moths emerging. (Crop Pest Cont. Mess.). PENNSYLVANIA - First entrance into apples in Franklin County. (Pepper). MARYLAND - Entries in apples May 23; moth emergence light. (U. Md.). DELAWARE - Peak emergence during week. (Late News). INDIANA - Egg laying and hatch slowed. Hatch expected to increase with warm weather. (Hamilton). Entries becoming noticeable in Orleans area. (Marshall). OHIO - First entry noticed on unsprayed apples June 3; cage emergence 90 percent complete. (C. R. Neiswander). ILLINOIS - About as many moths have emerged in past week as in preceding week; hatch expected to continue into June in Carbondale area. (Chandler). MISSOURI - Number of successful entries low, control excellent. (Martin). OREGON - Adults emerged in the Milton-Freewater area May 22 (Wallace); first adult taken at Hood River May 31. (Ellertson). WASHINGTON - Retarded by cool weather. (Luce).

CURCULIOS - OHIO - Very severe on unsprayed apples. (Cutright).

PLUM CURCULIO (Conotrachelus nenuphar) - MASSACHUSETTS - Activity slowed; late injury expected with high temperatures. (Crop Pest Cont. Mess.). NEW YORK - Activity light but persistent in Monroe County. (Corey). ILLINOIS - Gradual decrease. (Chandler).

RED-BANDED LEAF ROLLER (Argyrotaenia velutinana) - NEW YORK - Still active in several counties. (Wkly. News Lett.). VIRGINIA - Larvae of first new generation beginning to pupate. (Hill, May 27). MARYLAND - Pupae observed May 25. (U. Md.). INDIANA - First brood more numerous than previous two seasons in central area. Larvae active and pupating. (Hamilton).

APPLE MEALYBUG (Phenacoccus ac eris) - OREGON - Originally discovered in Oregon in 1951 in Brooks vicinity of Marion County, this insect now occurs over area 2 miles in diameter. It has been found only on filbert trees. Recent egg-mass counts average 28.4 per 2 feet length of branch on separate trees. Counts made on more heavily infested limbs 2-5 inches in diameter. (Roth).

CALIFORNIA PRIONUS (Prionus californicus) - NEW MEXICO - Appearing fairly abundantly at lights in the Mesilla Valley. (Ins. Lett., May 28).

ORIENTAL FRUIT MOTH (Grapholitha molesta) - INDIANA - Twig injury high in numerous plantings; most larvae have left twigs. Second-brood attack expected heavy where heavy twig injury occurred. ILLINOIS - Very few on peaches. (Chandler). MISSOURI - Injury noted in young peach orchards in Waverly area. (Wkly. Rept. Fr. Grs.).

PEACH TWIG BORER (Anarsia lineatella) - WASHINGTON - Active in orchards of Yakima Valley where no dormant spray used. (Luce).

LEAF MINERS - MARYLAND - Increasing in some orchards in Hancock area. (U. Md., Ent. Dept.). Unspotted tentiform leaf miner (Callisto geminatella) - Adults of this year's first generation began emerging May 24; present in large numbers, May 27. (Hill).

ORCHARD MITES - NEW YORK - European red mite summer eggs beginning to hatch in Columbia County. (Poray). VIRGINIA - European red mite more prevalent than normal in localized situations in orchards of northern Virginia. (Hill). INDIANA - Mite counts per 100 leaves on three varieties of apples averaged 17 on May 31 as compared to 196 per 100 leaves on June 1, 1954; figures taken on trees with no dormant spray. (Marshall). OHIO - European red mite more abundant than last year; many second generation mature and ovipositing. (C. R. Neiswander). UTAH - Clover mite (Bryobia praetiosa) infestations moderate to severe in many apple and peach orchards of northern area where control omitted. (Knowlton). OREGON - European red mite (Metatetranychus ulmi) beginning egg deposition on apples and pears May 28 at Hood River. (Ellertson).

CHERRY FRUIT FLY (Rhagoletis cingulata - WASHINGTON - Retarded by cool weather. (Luce). MICHIGAN - Numerous in Grand Rapids area. (Hutson).

EASTERN TENT CATERPILLAR (Malacosoma americanum) - WISCONSIN-Defoliated wild cherry and neglected orchards in southern area. (Chambers).

APHIDS - INDIANA - Subsiding in activity in Orleans area. (Marshall). ILLINOIS - In Carbondale area green apple aphid still moderate; rosy apple aphid leaving apples. (Chandler). UTAH - Green peach aphid (Myzus persicae) moderately to severely abundant and curling foliage in some peach orchards in Washington, Weber, and Kane Counties. (Knowlton, Burningham). Black cherry aphid (Myzus cerasi) injurious in s me orchards in Box Elder, Weber, Salt Lake and Washington Counties. (Knowlton). WASHINGTON - Mealy plum aphid (Hyalopterus arundinis) curling leaves of plum and requiring control measures at Parker. (Landis, Schopp). PENNSYLVANIA - Apple aphid (Aphis pomi) building up in orchards in Adams County. (Asquith). Black cherry aphid (Myzus cerasi) heavy on unsprayed sweet cherries in Centre County. (Adams).

Citrus Insect Conditions in Florida for Fourth Week in May, 1955

Increased rate of activity and hatching of PURPLE SCALE (Lepidosaphes beckii); 97 percent of groves inspected infested; highest activity was in West Coast and Brooksville areas. FLORIDA RED SCALE (Chrysomphalus aonidum) declining with increased activity and hatching expected in about a week; 57 percent of groves infested. Increased activity of CITRUS RED MITE (Metatetranychus citri) with 75 percent of groves infested; further increase expected. With 52 percent of groves infested, CITRUS RUST MITE (Phyllocoptruta oleivora) increased activity and rapid build-up expected; highest activity in West Coast and Indian River districts. (Pratt, Thompson and Johnson).

Mediterranean Fruit Fly Reported from Costa Rica

The Mediterranean fruit fly (Ceratitis capitata) has been found established on the central plateau of Costa Rica in an area 30 x 70 kilometers. The present serious situation may increase when additional hosts develop during rainy season and when coffee berries develop on 80,000 acres. Infestation in peaches heavy, moderate in oranges and other citrus. Adults observed frequently May 26. (Christensen, Stone). This insect is not known to occur in the continental United States. It was discovered in Florida in 1929 but was successfully eradicated, and no specimens have been found there since July 1930.

Mexican Fruit Fly Suppression Project in California

Suppressive operations on all host trees in California within five miles of the Mexican border were initiated in April, 1954, following discovery of this pest in Tijuana, Baja California, Mexico, January 1954. Since inception of operations, 282,717 host trees have been sprayed, using 61,825 gallons of bait material. No flies have been trapped in the California area of treatment except the single specimen taken in San Ysidro in August 1954. Plans are made to continue protective measures through this season and through the summer of 1956. (Armitage). According to available records, no Mexican fruit fly has been trapped in the northwest Mexico area since 2 specimens were taken at Tijuana November 23, 1954.

Citrus Blackfly Found at Brownsville, Texas

The following statement was released June 3 by Plant Pest Control Branch to State Plant Quarantine Officials:

This will advise you that an incipient infestation of citrus blackfly was discovered in the United States in the vicinity of Brownsville, Texas, on May 31, 1955. This infestation was found on a single lime tree on the grounds of a tourist court located outside of the center of town. No commercial groves occur in this area. Egg spirals and unemerged pupae were present on two leaves of this tree. No emerged pupal cases were found.

An intensive survey now underway has so far failed to reveal any additional infested trees in the vicinity. Appropriate officials in the

State of Texas are aware of this discovery and an aggressive cooperative spraying program is in progress to eradicate this infestation.

WALNUT APHID (Chromaphis juglandicola) - CALIFORNIA - Heavy
infestation over Stanislaus County. (Cal. Coop. Rpt.).

PECAN APHIDS - ILLINOIS - Moderate with considerable honeydew on
foliage in southern area. (Chandler).

TRUCK CROP INSECTS

COLORADO POTATO BEETLE (Leptinotarsa decemlineata) - RHODE
ISLAND - Eggs found on a few plants. (Kantack). DELAWARE -
Injuring potatoes near Dover, Lincoln and Georgetown. Destroying
tomatoes in Georgetown area. (Milliron). MARYLAND - All stages
present on unsprayed potatoes and tomatoes. (U. Md., Ent. Dept.).
VIRGINIA - Numerous enough to require treatment in Norfolk area and
Eastern Shore; attacking tomatoes on Eastern Shore. (Morris, May 27).
NORTH CAROLINA - Causing much damage to Irish potatoes and egg
plant in Duplin County. (Brett).

POTATO APHID (Macrosiphum solanifolii) - DELAWARE - Generally
not heavy as usual on potatoes but abundant on tomatoes some areas.
(Milliron).

POTATO FLEA BEETLE (Epitrix cucumeris) - DELAWARE - Severe
on potatoes at Townsend. (Milliron). RHODE ISLAND - Populations
continue heavy in untreated fields. (Kantack).

POTATO LEAFHOPPER (Empoasca fabae) - DELAWARE - Present in
most untreated potatoes in southern Kent and Sussex Counties. Increase
in one acreage east of Dover. (Milliron).

POTATO PSYLLID (Paratrioza cockerelli) - COLORADO - First adults
taken in Mesa County, May 25. (Colo. Exp. Sta.). WYOMING -
u t   continues to increase on non-economic host in Goshen County.
(Wallis)on

CUTWORMS - WASHINGTON - Damaging potatoes near Brownstown.
(Landis, Schopp). VIRGINIA - Attacking cabbage Carroll County.
(Price, May 28).

EGGPLANT LACE BUG (Gargaphia solani) - NORTH CAROLINA -
Causing damage to eggplant in Duplin County. (Brett).

EUROPEAN EARWIG (Forficula auricularia) - WASHINGTON - First-
brood nymphs injuring potatoes and flowers at Union Gap. (Landis,
Schopp).

TUBER FLEA BEETLE (Epitrix tuberis) - WYOMING - Emergence from hibernation started; 5 per 100 sweeps on Lycium in Goshen County. (Wallis).

CELERY LOOPER (Anagrapha falcifera) - FLORIDA - Heavy infestations on celery in the Everglades area; about 40 acres average about a dozen loopers per plant. (Denmark).

A CELERY TORTRICID (Tortrix ivana) - FLORIDA - Considerable numbers of adults with light larval populations on celery at Belle Glade; indication of increasing populations. (Denmark).

A CELERY CUTWORM (Platysenta sutor) - FLORIDA - Decreased population on celery in the Everglades area, averaging less than one per plant. (Denmark).

SIX-SPOTTED LEAFHOPPER (Macrosteles fascifrons) - MINNESOTA - From 5-10 adults per 20 sweeps on carrots at Brooklyn Center. (Minn. Rpt. Ser.).

IMPORTED CABBAGEWORM (Pieris rapae) - DELAWARE - Serious injury in a commercial planting west of Dover. (Milliron). NORTH CAROLINA - Causing damage to cabbage in Duplin County. (Brett).

HARLEQUIN BUG (Murgantia histrionica) - NORTH CAROLINA - Infestations building up rapidly in Duplin County on turnip and radish. (Brett).

CABBAGE APHID (Brevicoryne brassicae) - CALIFORNIA - Increasing in most fields Whittier area. (Campbell). NEW YORK - Building up on early cabbage in Dutchess County. (O'Leary).

CABBAGE MAGGOT (Hylemya brassicae) - WISCONSIN - Abundant on cabbage and radish in southern area. (Chambers). NEW YORK - Injury severe on unprotected plantings in Niagara County. (Stevenson).

CABBAGE SEEDPOD WEEVIL (Ceutorhynchus assimilis) - IDAHO - Light to medium populations in rape seed fields; controls being used; populations lower than in previous years. (Manis).

SPINACH LEAF MINER (Pegomya hyoscyami) - DELAWARE - Damage common in spinach near Hartly. (Milliron).

SPINACH FLEA BEETLE (Disonycha xanthomelas) - DELAWARE - Injury in commercial spinach near Hartly. (Milliron).

ALFALFA WEBWORM (Loxosteges commixtalis) - WYOMING - Heavy flight of moths laying eggs on beets in Goshen County. (Robb).

FLEA BEETLES - NORTH DAKOTA - Adults causing injury in seeded gardens and transplants in Fargo area. (Goodfellow). UTAH - Causing moderate damage to sugar beets in some areas. (Knowlton).

SUGAR BEET ROOT MAGGOT (Tetanops myopaeformis) - COLORADO- Showing in scattered fields in Weld County; early stages; insecticides ineffective. (Colo. Exp. Sta.).

SUGAR BEET WIREWORM (Limonius californicus) - CALIFORNIA - Eighty percent of the turnips in an untreated field in Los Angeles County were damaged. (Campbell).

BEET LEAFHOPPER (Circulifer tenellus) - COLORADO - Reaching 0.7 adults per linear foot of beet row; some spots 1.0. (ARS, Colo. Exp. Sta.).

SWEETPOTATO FLEA BEETLE (Chaetocnema confinis) - MARYLAND - General on sweetpotatoes and doing some damage in Wicomico and Somerset Counties. (U. Md., Ent. Dept.).

TWO-SPOTTED SPIDER MITE (Tetranychus bimaculatus) - CALIFORNIA- Showing on lima beans in Orange and Los Angeles Counties; continues to build up on strawberries. (Campbell).

MEXICAN BEAN BEETLE (Epilachna varivestis) - DELAWARE - Adults feeding on snap and lima beans from Rising Sun southward. (Milliron). VIRGINIA - Adults beginning to lay eggs freely; some hatch- ing. (Morris, May 27). PENNSYLVANIA - First adult and eggs in Adams County. (Pepper). MARYLAND - Light numbers of adults on snap beans in the Salisbury area, Wicomico County. (U. Md., Ent. Dept.).

LESSER CORNSTALK BORER (Elasmopalpus lignosellus) - MISSISSIPPI - Causing damage to cowpeas in George County, May 18. (Bond).

PEA APHID (Macrosiphum pisi) - DELAWARE - Commercial peas at Middleton and Lincoln severely damaged. (Milliron). RHODE ISLAND - Heavy populations on garden peas in some areas with 20 to 30 aphids per plant in the Wakefield area. (Kantack). PENNSYLVANIA - Moderate infestation in commercial peas; some control used in Centre County. (Adams). WISCONSIN - Increasing but no serious damage. (Chambers). MINNESOTA - Pan counts ranged from 0.01 to 3.3 in southcentral. (Minn. Ins. Rpt. Ser.). UTAH - Common in canning pea fields in northern area. (Knowlton).

BEAN LEAF BEETLE (Cerotoma trifurcata) - DELAWARE - Injury
to snap beans, Sussex County, and to lima beans north of Ellendale.
(Milliron). MARYLAND - Heavy foliage damage by adults on snap beans
in Wicomico and Somerset Counties. (U. Md., Ent. Dept.). VIRGINIA -
Feeding on snap beans and blackeyed peas in King William County.
(Willey, May 27). NORTH CAROLINA - Moderate foliage damage to
snap and soybeans in Washington, Martin, Pitt, Franklin, and Wake
Counties. (Scott).

A LEAF MINER (Liriomyza sp.) - ARIZONA - General on cantaloup
in Salt River Valley, becoming severe in some fields. (Ariz. Coop.
Rpt.).

SPOTTED CUCUMBER BEETLE (Diabrotica duodecimpunctata howardi)-
DELAWARE - Attacking squash at Ellendale. (Milliron). OKLAHOMA -
Light in cucurbits in north central. (Walton). PENNSYLVANIA - Adults
feeding on bean foliage in Adams County. (Pepper).

STRIPED CUCUMBER BEETLE (Acalymma vittata) - RHODE ISLAND -
Heavy infestation on cucumbers at Portsmouth with 2-3 beetles per
plant. (Kantack). NEW YORK - Large numbers on cucurbits in Dutchess
County. (O'Leary). MARYLAND - Damaging young squash in Montgomery
County. (U. Md., Ent. Dept.). DELAWARE - Destructive to large
cucumber planting west of Dover. (Milliron). OKLAHOMA - Light
populations in northcentral area. (Walton).

ONION THRIPS (Thrips tabaci) - DELAWARE - Injuring onions in
Ellendale-Georgetown area. (Milliron). TEXAS - Causing heavy, widespread damage to tomatoes in Cherokee County. (Gaines).

ONION MAGGOT (Hylemya antiqua) - OREGON - Adults emerging in
the Lake Labish area; first maggots noted May 1. (Crowell).

SLUGS - WASHINGTON - Very damaging to home garden vegetables
and perennials in South Bend area. (Tidrick).

OMNIVOROUS LEAF TIER (Cnephasia longana) - OREGON - Larvae
on peas and in strawberry fields in Marion and Washington Counties.
(Hanna).

RASPBERRY CANE BORER (Oberea bimaculata) - UTAH - Causing
some damage in Weber County. (Burningham).

RASPBERRY SAWFLY (Monophadnoides geniculatus) - MINNESOTA -
Abundant in central area. (Minn. Rpt. Ser.).

RASPBERRY ROOT BORER (Bembecia marginata) - WISCONSIN -
Quite serious in raspberry regions. (Chambers).

SPITTLEBUGS - CONNECTICUT - Stunting strawberries; very abundant
in local areas. (Johnson). NEW YORK - Heavy infestations in strawberries
in several counties. (Wkly. News Lett.).

STRAWBERRY CROWN MOTH (Ramosia bibionipennis) - UTAH - Damaging numbers at Provo. (Knowlton, Barlow).

STRAWBERRY LEAF ROLLER (Ancylis comptana fragariae) - IDAHO -
High populations of adults in strawberries near Moscow; first evidence
of season. (Manis).

A STRAWBERRY SAWFLY (Empria ignota) - MINNESOTA - Damage
heavy on plantings in central Minnesota near Brainerd, Aitkin, and
Nisswa. (Minn. Rpt. Ser.).

GREEN PEACH APHID (Myzus persicae) - NORTH CAROLINA - Up to
50 percent of plants with colonies in scattered fields of tobacco in
Harnett, Cumberland, Robeson, and Columbus Counties, most with
light infestations. (Mitchell). Light infestations in Jackson, Pitt, and
Northampton Counties. (Scott).

HORNWORMS (Protoparce spp.) - NORTH CAROLINA - A survey of
six southeastern counties averaged one egg or larva per ten tobacco
plants. (Mitchell). Reported from Lee, Martin, Pitt, Franklin, Wake,
Johnston, and Yadkin Counties. (Scott). TEXAS - Light local infestations of P. quinquemaculata in Nacogdoches County. (Markwardt).

SEED-CORN MAGGOT (Hylemya cilicrura) - CONNECTICUT - A
number of tobacco fields damaged and required replanting in the
Connecticut valley. (Johnson).

FOUR-SPOTTED TREE CRICKET (Oecanthus nigricornis quadripunc-
tatus) - NORTH CAROLINA - A few nymphs and adults in tobacco
fields in Cumberland, Robeson, Columbus, and Sampson Counties
with some evidence of damage from egg laying. (Mitchell).

TOBACCO BUDWORMS (Heliothis spp.) - NORTH CAROLINA -
Twenty-five percent of plants infested in one 4-acre field in Wilson
County (Guthrie); a few in Harnett, Columbus, and Sampson Counties
(Mitchell); countywide in Lee and Harmon; and reported from Martin,
Pitt, Franklin, and Wake Counties. (Scott).

TOBACCO FLEA BEETLE (Epitrix hirtipennis) - NORTH CAROLINA -
Light in southeastern counties (Mitchell); some damage in Martin,
Washington, Pitt, Wake, Franklin, and Johnston Counties. (Scott).

VEGETABLE WEEVIL (Listroderes costirostris obliquus) - NORTH
CAROLINA - Attacking tobacco in Vance County; 23 adults found under
one plant. (Jones).

WHITE-FRINGED BEETLES (Graphognathus spp.) - NORTH CAROLINA -
Severe damage to a half-acre field of tobacco in Columbus County.
(Jones, Rabb and Guthrie).

FOREST, ORNAMENTAL AND SHADE TREE INSECTS

ELM LEAF BEETLE (Galerucella xanthomelaena) - RHODE ISLAND -
Populations light with little injury. (Kantack). DELAWARE - Hatching
continues. Larval damage conspicuous some areas. (Milliron).
MISSISSIPPI - Young stages defoliating large elms in Marshall County.
(Hutchins). ILLINOIS - Serious damage already on Chinese elms in
southern area. (Moore et al).

SMALLER EUROPEAN ELM BARK BEETLE (Scolytus multistriatus) -
WISCONSIN - Abundant in dead and dying elms in southeastern area.
(Chambers). OKLAHOMA - Several hundred elms of all sizes infested
at Platt National Park, Sulphur, during third week of March. Problem
serious as elms   major tree species in park. (South. For. Pest
Reptr.).

SOUTHERN PINE BEETLE (Dendroctonus frontalis) - ALABAMA -
Outbreak in northern area continuing. Most serious infestations in
Franklin, Marion, Lawrence, Winston, Talladega, Clay, Calhoun, and
Cleburne Counties. Control continued. (South. For. Pest Reptr.).

BARK BEETLES - Ips engraver beetles and black turpentine beetles
noted in LOUISIANA in DeSoto and Rapides Parishes; in ALABAMA
in Talladega and Covington Counties; and in east TEXAS. Increased
Ips activity in Yell, Polk, Grant, and Bradley Counties, ARKANSAS.
Although rains in northern ALABAMA and MISSISSIPPI may reduce
Ips broods, both black turpentine beetles and Ips beetles may be expect-
ed in dry areas, particularly in stands recently cut or severely burned.
Latter pest may become quite serious if drought persists. (South. For.
Pes Reptr.). WISCONSIN - More abundant than usual in central area.
(Chambers).

WHITE-PINE WEEVIL (Pissodes strobi) - WISCONSIN - Causing injury
to white and jack pine and Norway spruce in north central area.
(Chambers).

CALIFORNIA OAKWORM (Phryganidia californica) - CALIFORNIA -
Severe damage on live oaks in Marin County. (Cal. Coop. Rept.).

JACK PINE BUDWORM (Choristoneura pinus) - MINNESOTA - Third-instar larvae in Brainerd, Bimidji, and Park Rapids triangle; heavy near Bimidji on staminate flowers; larvae, 1-3 per shoot feeding in new shoots of jack pine in reproduction near Brainerd. (Minn. Ins. Rpt. Serv.).

PINE TORTOISE SCALE (Toumeyella numismaticum) - WISCONSIN - Continues abundant in several counties in northeastern area. (Chambers).

PINE SPITTLEBUG (Aphrophora parallela) - MINNESOTA - Masses and nymphs abundant on jack pine reproduction north of Wilton. (Minn. Ins. Rpt. Serv.). MARYLAND - Extremely heavy on young loblolly pine on Eastern Shore. (U. Md., Ent. Dept.).

INTRODUCED PINE SAWFLY (Diprion simile) - WISCONSIN - Feeding heavily in northwestern area, indicating spread. (Chambers).

LEAF-CUTTING ANTS - Caused more than the usual amount of damage to young pines in east TEXAS and central LOUISIANA. New colonies appeared in low areas normally too wet for the ants. This resulted in increased control costs prior to planting. (South. For. Pest Reptr.).

BIRCH LEAF MINER (Fenusa pusilla) - CONNECTICUT - Severe, causing browning and defoliation of gray birches in many sections. (Johnson).

FOREST TENT CATERPILLAR (Malacosoma disstria) - WISCONSIN - Defoliation of much of birch and poplar in northwest. (Chambers). MINNESOTA - Larvae nearly mature in Pine and Carlton Counties June 1. (Minn. Ins. Rept. Serv.).

NORWAY MAPLE APHID (Periphyllus lyropictus) - IDAHO - First winged adults found on maple in Moscow. (Manis).

GREEN-STRIPED MAPLEWORM (Anisota rubicunda) - KANSAS - Adults taken in light traps in Doniphan County May 13 and 18. Moths also found on maple trees in Brown County. (Matthew).

MAPLE BLADDER-GALL MITE (Vasates quadripedes) - RHODE ISLAND - Galls very abundant on maple leaves in Cranston area (Mathewson); moderate throughout State (Kantack).

FALL CANKERWORM (Alsophila pometaria) - VIRGINIA - Defoliating forest trees and shrubbery in Floyd County. (Morris, May 27).

A PLANT BUG (Neoborus illitus) - CALIFORNIA - Severe damage to new growth of ash in Marin County. (Cal. Coop. Rpt.).

SPRUCE BUDWORM (Choristoneura fumiferana) - MINNESOTA -
Fifth-instar larvae at Bimidji. (Minn. Ins. Rpt. Serv.).

UGLY-NEST CATERPILLAR (Archips cerasivorana) - PENNSYLVANIA -
Fairly common Potter County on wild cherry and chokeberry. (Adams).

A SCALE (Aonidiella taxus) - FLORIDA - Collected on Podocarpus for
the first time in Florida. (Denmark).

BLACK VINE WEEVIL (Brachyrhinus sulcatus) - RHODE ISLAND -
Light numbers on yew. (Mathewson). Moderate numbers with damage
especially on Taxus. (Kantack).

SPIDER MITES - NEBRASKA - Building up on juniper and spruce in
eastern area. (Andersen). PENNSYLVANIA - Considerable infestation
of Paratetranychus ununguis in some plantings of juniper in Westmoreland County. (Udine).

BAGWORM (Thryidopteryx ephemeraeformis) - DELAWARE -
Hatching from Frederica southward. (Milliron). ILLINOIS - Small
larvae feeding on foliage of trees in southern area. (Moore et al).

AZALEA LACE BUG (Stephanitis pyrioides) - DELAWARE - Nymphs
appearing in Wilmington-Newark area. (Milliron).

AZALEA LEAF MINER (Gracilaria azaleella) - MARYLAND - Damaging
azaleas in Prince Georges and Montgomery Counties. (U. Md., Ent.
Dept.).

HOLLY LEAF MINERS (Phytomyza sp.) - RHODE ISLAND - Moderate
numbers common throughout the State. (Kantack, Mathewson).
OREGON - P. ilicis infestation serious on planting of 25 trees in
Portland; adults mating June 1. (Roth).

JAPANESE BEETLE (Popillia japonica) - VIRGINIA - One adult taken
on rose June 5 in Fairfax County. First report of season for area.
(Gentry).

SLUGS - PENNSYLVANIA - Very abundant, severely injuring and
killing petunias and other flowers in Westmoreland County. (Udine).

NARCISSUS BULB FLY (Lampetia equestris) - WASHINGTON - First
adult noted in San Juan County May 1. (Baker).

RHODODENDRON LACE BUG (Stephanitis rhododendri) - PENNSYL-
VANIA - Considerable on ornamental plantings of rhododendron in
Westmoreland County. (Udine).

SCALE INSECTS - MINNESOTA - Heavy local infestations on elm, ash, oak, and other deciduous trees and shrubs. (Minn. Ins. Rpt. Serv.). NEBRASKA - Oystershell scale (Lepidosaphes ulmi) very abundant on lilac, crawler stage present. (Andersen).

IMPORTED WILLOW LEAF BEETLE (Plagiodera versicolora) - RHODE ISLAND - Six to eight adults per leaf in Cranston area (Mathewson); heavy in the Scituate area (Kantack).

EUROPEAN ELM SCALE (Gossyparia spuria) - WISCONSIN - More abundant than usual. (Chambers). NEBRASKA - Very abundant on elms in western area. (Andersen).

ELM BORER (Saperda tridentata) - NEBRASKA - General infestation of elms in Lincoln and Scottsbluff areas. In some areas of Lincoln, three-fourths of trees are infested to the point where a reinfestation would kill the trees. (Hamilton, Andersen).

COTTON INSECTS

BOLL WEEVIL (Anthonomus grandis) - TEXAS - Slow build-up gener-ally in lower Rio Grande Valley. (Deer). Survival in hibernation cages at Waco June 3 was 7.8 percent as compared with 2.5 at same time in 1954. Weevils were found in 18 of 20 fields at average of 75 per acre. For corresponding week 1954 average of 45 per acre in 44 fields. (Parencia et al). Overwintered weevils reported entering fields in increasing numbers in the east, central, northeast, and north central areas. (Davis, Martin). LOUISIANA - Average number of weevils per acre in fields examined in Tallulah was 131 as against 465 for the corresponding period in 1954. (Gaines et al). MISSISSIPPI - First weevils reported in delta counties ranged from 0-264 and averaged 52 per acre. (Merkl et al). SOUTH CAROLINA - Percent survival to June 3 in cages in Florence County was 4.24 compared to 0.14 for the same date in 1954. (Walker, Hopkins, Jernigan). TENNESSEE - Only one adult weevil found in fields in west Tennessee. (Locke). NORTH CAROLINA - Infestations in Scotland County ranged from 0 to 166 per acre. (Mistric).

BOLLWORMS - ARIZONA - Occasional egg on cotton in Pima County. (Ariz. Coop. Rpt.). TEXAS - Slight increase last week in lower Rio Grande Valley. (Deer). MISSISSIPPI - Some adults in cotton in delta counties. (Merkl et al). SOUTH CAROLINA - Eggs present in some fields in Florence County. (Walker, Hopkins, Jernigan). NORTH CAROLINA - Observed on cotton in Scotland County with 19 injured terminals and 2.5 larvae per 100 feet of row in three fields. (Mistric).

PINK BOLLWORM (Pectinophora gossypiella) - TEXAS - Bloom inspection in lower Rio Grande Valley indicates more general infestations than last year. (Davis, Martin, May 31).

APHIDS - ARIZONA - Decreasing. (Ariz. Coop. Rept). MISSOURI - Increasing in scattered spots in fields, generally; only a few leaves infested to point of curling. (Kyd, Thomas). TEXAS - Declining in lower Rio Grande Valley. (Deer). SOUTH CAROLINA - Very light infestations in most fields. (Walker, Hopkins, Jernigan).

FLEAHOPPERS - TEXAS - Infestation in McLennan and Falls Counties exceeded 25 per 100 terminals in 5 of 20 fields. During corresponding week last year infestation averaged 2.1 per 100 terminals. (Parencia et al). Found in destructive numbers in scattered fields of upper coastal, south central, east, central, and northeast areas. (Davis, Martin, May 31). LOUISIANA - Appearing at 5-10 per 100 feet of row in Bossier, Red River, and Natchitoches Parishes. (Oliver). MISSISSIPPI - Infestations range from light to medium in delta counties. Some small squares being blasted in older cotton south of Greenville. (Merkl et al). TENNESSEE - Nymphs and adults increasing. (Locke).

BROWN COTTON LEAFWORM (Acontia dacia) - TEXAS - Considerably decreased. (Davis, Martin, May 31).

GRASSHOPPERS - TENNESSEE - Heavier than usual around fields; possibility of damage later. (Locke).

SPIDER MITES - ARIZONA - Tetranychus atlanticus and T. bimaculatus appearing in scattered places in the Eloy area. TEXAS - Damaging numbers in lower Rio Grande Valley. (Deer).

THRIPS - ARIZONA - Increasing at Buckeye, 15-20 per plant; lower counts in other areas. (Ariz. Coop. Rpt.). TEXAS - Injurious infestations continue in untreated fields in McLennan and Falls Counties. (Parencia et al). Most destructive insect in east, central, northeast, and north central areas. (Davis, Martin, May 31). MISSISSIPPI - Light to heavy in all fields examined in delta counties, with damage still occurring in fields treated 5 times. (Merkl et al). SOUTH CAROLINA - Infestations ranged from light to heavy in the Piedmont section. (Walker, Hopkins, Jernigan). TENNESSEE - Decreasing from damaging numbers of last week. (Locke). NORTH CAROLINA - Infestations on cotton in Halifax, Cleveland, Union, and Scotland Counties ranged from 0.4 to 8.0 per plant, with average of 2.2. (Mistric).

WHITEFLIES - ARIZONA - Abundant and causing some concern. (Ariz. Coop. Rpt.).

INSECTS AFFECTING MAN AND ANIMALS

MOSQUITOES - SOUTH CAROLINA - Problem at Fort Jackson. (Everts).
MISSISSIPPI - Outbreak in one section of Forrest County. Aedes sticticus,
A. vexans, Psorophora ferox, and P. varipes involved. (Broome).
LOUISIANA - Aedes sp. infestation very severe in rice and sugarcane
belt. (Oliver).

HORN FLY (Siphona irritans) - LOUISIANA - Fifty to 500 per head on
50 cattle, East Baton Rouge. (Oliver).

CATTLE GRUBS - MONTANA - Hypoderma lineatum adults becoming
active in warmer sections. (Roemhild).

CONENOSE BUGS (Triatoma spp.) - FLORIDA - Three engorged adults
of T. sanquisuga on bedding material at Alachua. (Tissot, Hunter).
ARIZONA - T. probably uhleri abundant and annoying at San Manuel,
Pinal County. (Ariz. Coop. Rpt.).

HOUSE FLIES - SOUTH CAROLINA - Abundant at Clemson, May 25.
(Nettles). PENNSYLVANIA - Becoming abundant in homes and barns
in south central area. (Pepper). ARIZONA - Flies, mainly house flies,
average of 5 highest grill counts in 2 towns in Maricopa and Pinal
Counties: May 16-20 (29. 4), May 23-27 (58. 4). (Ariz. Coop. Rpt.).

A FILTER FLY - SOUTH CAROLINA - This psychodid a problem at
Fort Jackson. (Everts).

FALSE STABLE FLY (Muscina stabulans) - IDAHO - Two puparia
and one full-grown larva obtained from a hospital patient in Lewiston.
Specimens apparently passed in feces. According to M. T. James,
who made the determination, this is a common species in intestinal
myiasis. (Merkeley, Manis, May 28).

TICKS - MONTANA - Wood ticks abundant on sheep in northern Custer
and Rosebud Counties. Rocky Mountain wood tick abundant at lower
elevations in western part of State. (Roemhild). WISCONSIN - Derma-
centor variabilis abundant in north central counties. (Chambers).
VIRGINIA - Very heavy on a 40 to 50-acre pasture in Lunenburg
County. Control measures applied but cattle again covered with ticks
in a day or two. (Powers).

BENEFICIAL INSECTS

MANTIDS - NEW MEXICO - Noticeable numbers in shrubbery throughout southern half of State. (Ins. Lett., May 28).

LADY BEETLES - NEW MEXICO - Large numbers, principally Hippodamia convergens, throughout aphid-infested alfalfa fields; however, not controlling yellow clover aphid. (Ins. Lett., May 28).

STORED PRODUCTS INSECTS

Stored Grain Insects in Arkansas

During a survey of new grain going into storage on farms in east central area the predominant species found was Oryzaephilus surinamensis, followed by Rhyzopertha dominica, Sitophilus oryza, Sitotroga cerealella, Tribolium sp., Laemophloeus sp., and Tenebroides mauritanicus. Potential for new infestations on some farms. (Warren).

ANGOUMOIS GRAIN MOTH (Sitotroga cerealella) - LOUISIANA - Several cribs of corn over State have severe infestations. (Oliver).

RECENT INTERCEPTIONS AT PORTS OF ENTRY

Living adults of the Chinese rose beetle (Adoretus sinicus (Burm.)) were intercepted recently in airplane baggage and as a stowaway on airplanes during preflight inspection of aircraft leaving Hawaii for the mainland at Hickam Field and Honolulu airport, T. H. (Macdougall, Mason, Wakefield). This insect has been reported injurious to a variety of plants in Hawaii, including rose, grape, okra, string beans, sweet potatoes, canna, sugarcane, and others. Injury is due to the adults feeding on the foliage. Affected leaves are peppered with holes and more or less skeletonized. The adults also feed on the buds and flowers.

Observations on the bionomics of the Chinese rose beetle indicate the eggs are deposited in the soil. They hatch in about 4 days. The larvae develop in the soil and feed on decaying plant material. Pupation occurs in earthern cells in the soil. A life cycle may be completed in 6-7 weeks. Adults hide in leaves, plant debris, or in loose soil during the day, coming out to feed at night. A. sinicus is said to occur throughout southern Asia. It is believed to have been introduced into Hawaii sometime before 1896, probably with soil around the roots of plants. It has been intercepted on a number of occasions in airplane baggage, cargo and mail, and as a stowaway in planes leaving Hawaii for the mainland, and with cut flowers and plants from Hawaii at west coast ports. It is not known to occur in the continental United States.

(Compiled - Plant Quarantine Branch).

LIGHT TRAP COLLECTIONS

		Pseudal. unipun.	Prod. ornith.	Agrotis ypsilon	Perid. marq.	Feltia subter.	Heliothis armig.	Heliothis vires.	Protoparce sexta quin
TEXAS		3	7		1	9	8		
Waco	5/28-6/3	3	7		1	9	8		
LOUISIANA									
Bat. Rge.*	5/27-6/3	8	181	2		583	18		
Franklin	5/26-31		7		6	14	5		
Curtis	5/26-30	54	80	10	11	3	20		
Bunkie	5/23-6/1	5			6	2	26		
Tallulah*	5/27-6/3	52	158	10	35	30	49	3	4
ARKANSAS									
Hope	5/20-5/26	13		16	34		3		
Stuttgart	5/26-6/2	341		40	28		1		
Van Buren	5/19-6/2	58		42	23		10		
Varner	5/20-6/2	84		12	29		31		
Fayetteville	5/30-6/2	7		3	7		15		
Clarksville	5/26-6/2	47	54	32	14		6		52 sp.
MISSISSIPPI (Counties)									
Coahoma	5/27-6/3	49	9	3	9	1	3		
Humphreys		95	25	1	18	1	9		
Oktibbeha		71	167	8	28	9	13		
Pearl River		9	2				6		
Washington*	5/27-6/3	672	173	39	140	9	45	3	17 spp.

* Two traps at Baton Rouge, 3 at Tallulah, 2 in Washington County.

- Continued on page 528 -

LIGHT TRAP COLLECTIONS

		Pseudal. unipun.	Prod. ornith.	Agrotis ypsilon	Period. marg.	Feltia subter.	Heliothis armig.	vires.	Protoparce sexta	quin.
ALABAMA										
Auburn	5/28-6/3	1	12				4			
GEORGIA (Counties)										
Clarke	5/21-27	20	9			60	4			
Tift	5/22-28	2	5			4			5	
Spalding	5/21-27		1	1				9 sp.	3	
SO. CAROLINA (Counties)										
Oconee	5/29-6/4	36	5	2		4			9	4
Charleston	5/24-30		4	1	6	3		7 sp.	9	
NO. CAROLINA (County)										
Duplin	5/30-6/5		2	1					14	1
MARYLAND (County)										
Montgomery		4		3					2	
KANSAS										
Manhattan	5/23-6/3	331								
Hays	5/26-31	72								
Wathena	5/15-24	64								
COLORADO (County)										
Otero	5/22	110 (peak)					1 (first of season)			

Some other collections of interest: LOUISIANA (Tallulah) Loxostege similalis 2311; ARKANSAS (Stuttgart) Laphygma frugiperda 21 - first of season for area; GEORGIA (Spalding) Elasmopalpus lignosellus 1; SOUTH CAROLINA (Charleston) Conoderus vagus 1427; KANSAS (Manhattan) Chorizagrotis auxiliaris 52, (Hays) 53; COLORADO (Otero) Loxostege sticticalis 4000 May 20, 5000 May 23; C. auxiliaris 156 May 20, 225 May 22.

MISCELLANEOUS INSECTS

WHARF BORER (Nacerdes melanura) - NEW YORK - Heavy infestation
in underpinning of dwelling in the Bronx. Adults active. (Ramsey, P. C.
Branch). PENNSYLVANIA - Numbers in basement of house in Allegheny
County. (Udine).

BLACK CARPENTER ANT (Camponotus herculeanus pennsylvanicus) -
WISCONSIN - Unusually annoying to home owners throughout southern
part of State. (Chambers).

TERMITES - IDAHO - Reports from widely scattered areas on damage
from subterranean termites. Either infestations are increasing or
home owners are becoming conscious of them. (Portman, May 28).

ADDITIONAL NOTES

TENNESSEE - Light trap records for the week ending June 5:
Protoparce spp. - 196; P. quinquemaculata - 168; P. sexta - 52;
Euetheola rugiceps - 3000.

ARKANSAS - PEA APHID infestations on alfalfa increased in central
counties. Second growth alfalfa carrying populations up to 350 per
20 sweeps. Three-cornered alfalfa hopper causing some damage to
alfalfa in this area. (Warren).

TENNESSEE - ARMYWORM outbreak over. GRASSHOPPER nymphs
moderately abundant over State. SUGARCANE BEETLE continues to
cause serious damage to corn plantings, particularly in western area.
Considerable corn replanted. Cotton and strawberries also attacked
locally. BLISTER BEETLES numerous in scattered alfalfa fields
in middle Tennessee. Light infestations of first-instar EUROPEAN
CORN BORER in corn plantings across State. (Mullett).

IOWA - EUROPEAN CORN BORER as of June 4 - 100 percent pupation
and 40 percent emergence of moths in northwest Iowa. Moth flight
very heavy in central section with 55 egg masses per 100 plants on
20-inch corn. Northeast Iowa - no eggs found on 20-inch corn on
June 2. Cool nights slowed activity slightly. Anticipate hatching to
be general in central area by June 10 and in northern by June 18.
CORN LEAF APHIDS building up on corn. CORN FLEA BEETLES
damaging 8-inch corn in Delaware County. POTATO LEAFHOPPERS
in large numbers in central Iowa, June 1. GRASSHOPPERS damaging
soybeans in Page County, in one field at the rate of 10 per square foot.
First adults lesser migratory grasshoppers observed in southwest
Iowa. Majority in third to fifth instar. (Harris).

MINNESOTA - EUROPEAN CORN BORER pupation and moth emergence advanced sharply in south-central and southwest districts with pupation almost complete (27 percent moth emergence). Reports from east-central district indicate about 30 percent pupation and 1-2 percent moth emergence for same period. Corn heights in relation to borer development very low and egg deposition, already begun, should be of little consequence in corn for the next week or 10 days. GRASSHOPPERS beginning to cause some crop damage in southeastern, central, and northwestern areas. In northwestern area Melanoplus bivittatus and M. mexicanus hatch practically complete. Warm weather has promoted high survival in general. ARMYWORMS - Adults continue to appear in light traps. (Minn. Ins. Rpt. Serv.).

MISSOURI - GRASSHOPPERS - Hatch of Melanoplus differentialis continues over northern half of State. Hatch of other species practically complete. Infestations general in southwest, northwest, and central areas. Overall situation is considered severe to critical in all areas of State. (Kyd, Thomas).

WYOMING - GRASSHOPPERS developing in some range areas of Johnson, Goshen, Campbell, and Platte Counties. (Spackman).

VIRGINIA - SEED-CORN MAGGOTS and SOUTHERN CORN ROOTWORMS causing moderate to severe damage to corn in Pittsylvania County. (Dominick). Next brood of RED-BANDED LEAF ROLLERS expected to be heavy in several apple orchards in Rappahannock County. (Lyne). HORNWORM adults found in 6 light traps in Pittsylvania County for the week ending May 28 were Protoparce quinquemaculata 894 and P. sexta 186. (Dominick). ARMYWORM infestations continue to be reported but severe damage not expected. (Morris).

Coop
Unite
CAT7
U.S.1

[74]
Dec 1

6 pe
United
CAT7
US D

[73]
Dec 1